U0108639

飲食同盟會 編著

新手入廚系列

蝦蟹滋味

前言

所謂靠山吃山、靠水吃水，香港人住海邊，自然好吃海鮮，對吃海鮮也很有心得，例如何時該吃魚、何時該吃蝦、怎樣搭配才好吃、怎樣煮才有營養，相信都不會難倒香港人。而當中大受歡迎的，要數蝦蟹。

蝦蟹種類很多，烹煮容易，尤其是蝦，無論作為主菜還是配菜，都很討好，於是蝦成了飯桌上的常菜。蟹的種類雖不如蝦多，但卻很受愛好美食的朋友重視。

作為一個國際都會，香港人的飲食也大受五湖四海的影響，以至於吃蝦蟹，也吃得五花八門、琳琅滿目：中式的煎炒，西式的涼拌、泰式湯煮、日式油炸……數之不盡。

正是出於對蝦蟹的喜愛，本書特選了四十多款不同風味、較受歡迎的蝦蟹菜式。由於本書是針對家庭讀者的，因此這些菜煮法會較簡單，而且搭配靈活多變，讓各位能用簡單的方法、用具，做出不一樣的菜，在家裏也能吃到好吃又味道多變的蝦蟹菜式。

目錄

蝦技法

選蝦的竅門

1. 新鮮的蝦，頭部完整，身體仍保持原有的彎曲度。
2. 活蝦會活蹦亂跳，色澤青綠或青白，殼發亮。
3. 相反，變壞的蝦頭殼容易脫落，肉質鬆軟，色澤赤紅。

剪蝦

為了要大蝦入味、食用方便又美觀，要先進行「剪蝦」工作：用手托起大蝦，蝦背向上，用剪刀剪去鬚和槍。

蝦膠

⦿ 材料

蝦肉 600 克

⦿ 配料

肥豬肉 75 克
馬蹄粒適量
陳皮茸適量

⦿ 醃料

鹽 1 茶匙
澄麵 4 克
生粉 4 克
麻油 1/2 茶匙
雞蛋白 1 隻
胡椒粉少許

⦿ 做法 | Method

1. 鮮蝦去殼。
2. 在盆裏放入清水和 1/2 茶匙鹽，放入鮮蝦泡 5 分鐘。撈起鮮蝦，吸乾水分，用刀把蝦肉拍爛，和配料一起混和，稍剁。
3. 加調味料，混合後拍打至凝固成膠狀，放進冰箱冷藏一、兩小時。

蟹技法

1. 新鮮的螃蟹，蟹腳活躍堅硬，若反應遲鈍、又有脫落的話，可能已變壞。
2. 可用手掂一掂重量，或捏捏蟹殼，殼硬表示肉厚實，肥滿，軟殼者肉少、鬆軟。
3. 分辨雌蟹和雄蟹：雄蟹肚臍尖長、肉質螯肥；雌蟹肚臍呈圓形，內有金黃香頤的蟹黃。

蟹的處理方法

1. 先把蟹泡在冰水裏，讓其死去停止活動。
2. 打開蟹蓋，如果有蟹臍，要先去除蟹臍。
3. 除去鰓和臍，洗淨。
4. 用刀背拍破蟹鉗。
5. 將蟹剁成塊。

防止食蟹中毒法

蟹性涼、味甘鹹，帶有小毒，所以必須選購新鮮螃蟹，並經過仔細處理才行。而蟹黃和蟹油不易消化，因此體質寒涼者和胃弱者不宜多吃。即使吃了，也應在用完蟹後喝點薑湯或桂圓湯，以便驅寒。

處理蟹要留心

胃：胃在螃蟹眼部內側，狀如沙囊，裏面儲存了螃蟹吃剩的食物殘渣，若誤食會引起胃痛、上吐下瀉等徵狀。

肺：位於螃蟹下身部位的兩側，灰色、狀如菊花瓣，必須清除乾淨。

腸：打開蟹的腹蓋，會看到腹部末端—即是肛門，而連接肛門的就是腸子，要徹底清除好，以免中毒。

心：位於兩肺中間，也是不能誤食的。

材料 | Ingredients

素翅 100 克
瘦豬肉 100 克
蟹肉 50 克
雞湯 2 杯
清水 4 杯

100g imitation shark-fin
100g lean pork
50g crab meat
2 cups chicken broth
4 cups water

4~6 人
Serves 4~6

40~45 分鐘
40~45 minutes

芡汁 | Thickening

老抽 1 湯匙
生粉 2 湯匙
清水 3 湯匙
鹽 1/2 茶匙

1 tbsp dark soy sauce
2 tbsps caltrop starch
3 tbsps water
1/2 tsp salt

調味料 | Seasonings

麻油 1 茶匙
胡椒粉少許

1 tsp sesame oil
Dash of pepper

入廚貼士 | Cooking Tips

- 可把煲完老火湯的瘦豬肉撕成肉絲，效果相若，這樣做不但能善用資源，又可以省錢，一舉兩得。

- You may use cooked lean pork from soup, the texture is similar.

做法 | Method

1. 把雞湯、清水和瘦豬肉煲 30 分鐘，取出，把瘦豬肉撕成絲。
2. 把瘦豬肉絲、素翅一起回放煲中，以上湯煮滾。
3. 慢慢加入芡汁料，期間不停攪拌至湯稠，熄火。
4. 下蟹肉拌勻，加入調味拌勻，即可享用。

1. Boil chicken broth, lean pork and water for 30 minutes, remove and tear lean pork into shreds.
2. Place shredded lean pork together with imitation shark-fin back into the saucepan and bring to the boil with the broth.
3. Add thickening in gradually, stir constantly until it has thickened and remove from heat.
4. Add crab meat in and mix well, then add seasonings and serve.

4~6 人
Serves 4~6

20~25 分鐘
20~25 minutes

Tom Yum Goong

冬蔭公海鮮湯

材料 | Ingredients

雜海鮮 600 克	600g mixed seafood
焗香蝦頭 300 克	300g baked shrimp heads
冬蔭公湯 1 包	1 pack Tom Yum Goong soup base
上湯 1 1/2 公升	1 1/2 litres chicken broth
香茅 1 枝	1 stick lemongrass
九層塔 2 棵	2 sprigs thai basils
辣椒 3-4 隻	3-4 red chilies
辣椒膏 1 湯匙	1 tbsp chili paste
蒜茸 1 茶匙	1 tsp minced garlic

調味料 | Seasonings

鹽 1/2 茶匙	1/2 tsp salt

做法 | Method

1. 雜海鮮清洗乾淨，汆水過冷，備用。
2. 香茅切碎；九層塔洗淨，留葉；辣椒洗淨切塊。
3. 鑊燒熱，下油爆香蒜茸、辣椒、辣椒膏和香茅，放下湯包、已焗香的蝦頭和上湯煮 15-20 分鐘，隔渣留湯。
4. 雜海鮮放入湯底煮滾，加入九層塔葉煮滾，下調味料即成。

1. Wash mixed seafood. Blanch. Rinse in cold water.
2. Chop lemongrass. Wash Thai basils. Retain the leaves only. Wash red chili. Cut into wedges.
3. Heat some oil in wok. Sauté the garlic, red chili, chili paste and lemongrass until aromatic. Add Tom Yum Gong soup base, baked shrimp heads and chicken broth. Cook for 15-20 minutes. Remove the residues.
4. Add mixed vegetables into Tom Yum Gong broth. Add Thai basil leaves. Bring to a boil. Finally add seasonings.

海鮮冬茸羹

Mashed Winter Melon and Seafood Thick Soup

材料 | Ingredients

冬瓜 600 克	600g winter melon
中蝦 3 隻	3 pcs medium shrimp
帶子 2 隻	2 pcs scallop
鮮魷 100 克	100g squid
瑤柱 3 粒	3 cloves dried scallop
雞蛋白 2 隻	2 egg white
薑 2 片	2 slices ginger
雞湯 2 杯	2 cups chicken broth
清水 1 杯	1 cup water

醃料 | Marinade

生粉 1 茶匙	1 tsp caltrop starch
鹽 1/2 茶匙	1/2 tsp salt
糖 1/4 茶匙	1/4 tsp sugar
胡椒粉少許	Pinch of pepper

調味料 | Seasonings

胡椒粉、鹽各適量

Some salt and pepper

荚汁 | Thickening

生粉 2 湯匙
清水 3 湯匙

2 tbsps caltrop starch
3 tbsps water

做法 | Method

1. 中蝦、帶子和鮮魷全部切粒，下醃料醃 15 分鐘。
2. 冬瓜切件；瑤柱浸軟；雞蛋白打勻。
3. 將雞湯、瑤柱和清水煮滾，放入冬瓜件煮至再次滾起熄火，焗 15-25 分鐘，取出，用餐匙將瓜肉刮成茸。
4. 下 1 湯匙油爆香薑片，中蝦、帶子、鮮魷下鑊略炒，再與冬瓜茸一起放回上湯中煮滾。
5. 慢慢加入生粉水，然後不停攪拌至湯稠，熄火。
6. 沿邊下雞蛋白，攪拌數下，加入調味即可享用。

1. Cut shrimps, scallops and squids into dice and marinate for 15 minutes.
2. Cut winter melon into pieces, soak dried scallops until soft and beat egg whites.
3. Bring chicken broth, dried scallops and water to the boil. Add winter melon in and when it bring to the boil again, remove from heat and cover with a lid for 15–25 minutes. Remove and mash winter melon with a spoon.
4. Heat 1 tbsp of oil and sauté ginger until aromatic. Stir-fry shrimps, scallops and squids for a while, put them together with the mashed winter melon into the chicken broth and bring to the boil.
5. Add caltrop starch gradually, stir constantly until the soup is thickened and remove from heat.
6. Slide egg whites in, stir for a while, add seasonings and serve.

番茄海鮮湯

Tomato Seafood Soup

4~6 人
Serves 4~6

25~30 分鐘
25~30 minutes

材料 | Ingredients

蟹 2 隻
蜆 600 克
番茄 600 克
芫荽頭 3–5 棵
芹菜 2–3 棵
薯仔 1 個
上湯 1 公升

2 crabs
600g clams
600g tomatoes
3–5 sprigs parsley root
2–3 sprigs celery
1 potato
1 litre chicken broth

調味料 | Seasonings

鹽適量
胡椒粉適量

Pinch of salt
Pinch of pepper

做法 | Method

1. 蟹劏洗乾淨，切件。
2. 薯仔去皮切角；番茄浸熱水片刻，去皮，去核，切角。
3. 芹菜洗淨，切段。
4. 把上湯煲滾，放入所有材料滾 25–30 分鐘，下調味，上桌。

1. Wash crabs. Chop.
2. Peel potato. Cut into wedges. Soak tomatoes in hot water for a while. Peel and remove the seeds. Cut into wedges.
3. Wash celery and section.
4. Bring chicken broth to a boil. Add all ingredients. Cook for 25–30 minutes. Add seasonings. Dish.

酸甜雲吞

Sweet and Sour Wontons

◯◯◯ 材料 | Ingredients

春卷皮 1 包	1 packet spring roll wrapping
蝦膠 300 克	300g minced shrimps
菠蘿粒 100 克	100g pineapple dice
番茄 1 個（切粒）	1 tomato (diced)
三色椒粒 40 克	40g bell peppers of three colours (diced)
蒜茸 1 茶匙	1 tsp minced garlic

◯◯◯ 調味料 | Seasonings

雞蛋白 1 隻	1 egg white
鹽 1 茶匙	1 tsp salt
糖 1/2 茶匙	1/2 tsp sugar
生粉 2 茶匙	2 tsps caltrop starch
胡椒粉適量	Some pepper
麻油適量	Some sesame seed oil

4~6 人
Serves 4~6

10~15 分鐘
10~15 minutes

⟨⟩ 酸甜汁 │ Sweet and Sour Sauce

茄汁 1/3 杯	1/3 cup ketchup
白醋 1/4 杯	1/4 cup white vinegar
片糖 1 片	1 pc cane sugar
鹽 1/3 茶匙	1/3 tsp salt
日式燒汁 1 湯匙	1 tbsp teriyaki sauce
老抽 1 茶匙	1 tsp dark soy sauce
生粉 1 茶匙	1 tsp caltrop starch
水 50 毫升	50ml water

⟨⟩ 做法 │ Method

1. 蝦膠加入調味料拌勻。
2. 春卷皮解凍後，用 2 片春卷皮重疊，包上少許蝦膠成雲吞狀，放入八成滾油中炸至金黃，盛起瀝油。
3. 熱鑊爆白蒜茸，加入三色椒炒香，盛起。酸甜汁拌勻煮滾，加入三色椒和菠蘿拌勻，沾雲吞供食。

1. Add seasonings to minced shrimps and stir until sticky.
2. Defrost spring roll wrapping, stack two pieces together, wrap in some shrimp paste and fold as a wonton. Deep-fry wontons in hot oil until golden. Strain excess oil.
3. Stir-fry minced garlic in a heated pan, add in bell peppers, stir-fry thoroughly, set aside. Cook mixed sweet and sour sauce until it boils, add in bell peppers and pineapple, and mix well. Serve with wontons.

入廚貼士 | Cooking Tips
- 用春卷皮做雲吞，是傳統酒樓的雲吞，香脆而沒有鹼水味。
- It is a traditional method of making wontons with spring roll pastry adopted by restaurants. The pastry makes wontons crispy without alkaline taste.

錦鹵蛋散

Sweet and Sour Crisp Egg Twists

材料 | Ingredients

上海餛飩皮 150 克
菠蘿粒 100 克
車厘茄 50 克
青、紅椒各 50 克
乾葱頭 50 克
蒜茸 1 茶匙

150g Shanghai dumpling wrapping
100g pineapple dices
50g cherry tomatoes
50g green bell pepper
50g red bell pepper
50g shallots
1 tsp minced garlic

◯◯ 茄汁 | Tomato Sauce

茄汁 1/3 杯	1/3 cup ketchup
白醋 50 毫升	50ml white vinegar
照燒汁 50 毫升	50ml terriyaki sauce
黃糖 5 湯匙	5 tbsps brown sugar
老抽 2 茶匙	2 tsps dark soy sauce
鹽 1/2 茶匙	1/2 tsp salt
水 50 毫升	50ml water

◯◯ 做法 | Method

1. 車厘茄、青甜椒、紅甜椒和乾蔥頭洗淨切粒,備用。

2. 茄汁料拌勻,熱鑊下油爆香蒜茸,放入乾蔥頭,青、紅椒和車厘茄,加入茄汁料煮滾,放入菠蘿粒兜勻。

3. 把上海餛飩皮剪成長條,2-3 片重疊一起,放熱油中炸至金黃,取出瀝油,上碟,伴以茄汁享用。

1. Wash cherry tomatoes, green bell pepper, red bell pepper and shallots, then cut them into dice. Set aside.

2. Mix tomato sauce ingredients thoroughly. Put some oil in a heated frying-pan, add minced garlic and sauté until aromatic. Add shallots, green and red bell peppers and cherry tomatoes, pour in the tomato sauce ingredients and bring it to a boil. Add pineapple dice and stir thoroughly.

3. Cut dumpling wrappings into long pieces, stack 2 or 3 slices together and deep-fry in hot oil until golden. Drain excess oil and then place on a plate. Serve with the tomato sauce.

蔗蝦

Shrimp with Sugar Cane

材料 | Ingredients

蝦膠 450 克
竹蔗 4–6 枝

450g minced shrimps
4–6 sugar canes

4~6 人
Serves 4~6

12~15 分鐘
12~15 minutes

調味料 | Seasonings

生粉 2 茶匙
砂糖 1 茶匙
鹽 1/2 茶匙
雞蛋白 1 隻
胡椒粉少許

2 tsp caltrop starch
1 tsp sugar
1/2 tsp salt
1 egg white
Some pepper

伴食料 | Side Dish

青檸（切角）1 個
泰式雞醬適量

1 lime (cut into wedges)
Some Thai chicken sauce

做法 | Method

1. 蝦膠加入調味料拌勻，攪至起膠，放冰箱中冷藏 30 分鐘。

2. 在竹蔗首末兩端的中間位置削一凹位，洗淨抹乾。

3. 用手塗油，把蝦膠慢慢套在竹蔗上弄成棒球棒般。

4. 燒一鍋新鮮油至八成滾，放入蔗蝦炸至金黃，取出瀝油，與伴食料享用。

1. Mix minced shrimps with seasonings until it becomes very sticky. Chill in a refrigerator for 30 minutes.

2. Cut a hole at both ends of a sugar cane. Wash and pat dry.

3. Rub some oil on the sugar cane, put some minced shrimp on it slowly and make into the shape of a baseball bat.

4. Heat a wok of oil until it becomes fairly hot. Put in shrimps on sugar canes and deep-fry until golden. Drain excess oil. Serve with lime or Thai chicken sauce.

大蝦蘋果沙律

Shrimp and Apple Salad

◯◯◯ 材料 | Ingredients

蝦 150 克
紅蘋果 1 個
青蘋果 1 個
檸檬汁少許

150g shrimp
1 red apple
1 green apple
Pinch of lemon juice

◎◎ 沙律汁 | Dressing

蛋黃醬 3 湯匙
煉奶 1 茶匙

3 tbsps mayonnaise
1 tsp concentrated milk

◎◎ 做法 | Method

1. 蝦挑去腸、洗淨，以大火蒸熟，浸冰水，去殼。
2. 青、紅蘋果切粒，與檸檬汁拌勻。
3. 沙律汁拌勻，撈入蘋果粒，鋪上蝦即成。

1. Wash shrimps. Remove the intestines. Steam over high heat until done. Soak in iced water. Remove the shells.
2. Dice the green and red apple. Combine with lemon juice and toss well.
3. Add salad dressing. Toss well. Top with shrimps.

咖喱炒蟹

Stir-fried Curry Crabs

⟨⟨⟨⟩⟩⟩ 材料 | Ingredients

蟹 2 隻
薯仔塊 150 克（炸透）
洋葱塊 80 克
青、紅椒段各 80 克
中國芹菜段 80 克
辣椒 2-3 隻（切段）

油咖喱 2 湯匙
蒜茸、薑茸 1 湯匙
乾葱茸 1 湯匙
咖喱粉 1 茶匙
紅椒粉 1/4 茶匙

2 crabs
150g deep-fried wedged potatoes
80g stripped onion
80g stripped red bell peppers
80g stripped green bell peppers
80g stripped Chinese celery
2-3 stripped red chilies
2 tbsps curry oil
1 tbsp minced garlic
1 tbsp minced ginger
1 tbsp minced dried shallot
1 tsp curry powder
1/4 tsp paprika

4~6 人
Serves 4~6

20~25 分鐘
20~25 minutes

調味料 | Seasonings

上湯 1 杯
茄汁 1 湯匙
糖 2 茶匙
鹽 1/2 茶匙
胡椒粉適量

1 cup chicken broth
1 tbsp ketchup
2 tsps white sugar
1/2 tsp salt
Pinch of pepper

芡汁 | Thickening

生粉 2 茶匙
清水 6 湯匙

2 tsps caltrop starch
6 tbsps water

做法 | Method

1. 將蟹洗乾淨，切件，加醃料醃 5 分鐘。

2. 鑊燒熱，放 5–6 湯匙油，分別爆香青椒、紅椒、洋葱、芹菜，取起留用。

3. 原鑊炒香所有香料，放入蟹件煎至出香味，灒酒，加入調味料和薯仔件拌勻，加蓋焗 15–20 分鐘。放入雜蔬拌勻，加入芡汁料煮稠，再焗 5 分鐘或至熟透。

1. Wash crabs. Chop. Marinate for 5 minutes.
2. Heat 5–6 tbsps of oil in wok. Sauté green bell peppers, red bell peppers, onion and Chinese celery separately until aromatic.
3. Sauté spices until aromatic. Add crabs. Sauté until aromatic. Sprinkle with some cooking wine. Add seasonings and potatoes. Sauté well. Cover. Cook for 15–20 minutes. Add mixed vegetables. Sauté well. Add thickening. Cook for further 5 minutes or until done.

黃金蝦

Fried Shrimps with Salted Egg Yolk

⃝⃝⃝ 材料 | Ingredients

中蝦 400 克
鹹蛋黃 3 個

400g medium shrimps
3 pcs salted egg yolk

炸漿料 | Deep-fry Batter

粟粉 1/2 杯

1/2 cup caltrop starch

調味料 | Seasonings

鹽少許

A pinch of salt

做法 | Method

1. 蝦修剪好，去腸，洗淨，瀝乾。
2. 鹹蛋黃蒸熟，壓碎，備用。
3. 每隻蝦逐一撲上粟粉，輕輕放入滾油中，炸至金黃色，取出瀝油。
4. 把油倒掉，只在鍋中留下 2 湯匙油，用中火兜炒鹹蛋黃碎，再加入蝦回鑊兜勻，加入少許鹽，即可上碟。

1. Trim and devein shrimps. Wash and drain.
2. Steam salted egg yolk until cooked, crush and set aside.
3. Coat each of the shrimp with caltrop starch, deep-fry in hot oil until golden and drain.
4. Pour oil out and leaving 2 tbsps of oil in wok, stir-fry salted egg over medium heat, toss with shrimps until well combined add a pinch of salt and dish up.

蒜茸玉簪蝦

Garlic Shrimps in Tomato Sauce

⊂⊃ 材料 | Ingredients

海中蝦 450 克
葱 2–3 條（切絲）
茄汁 1/3 杯
蒜茸 2 湯匙
辣椒碎 1 湯匙
薑茸 2 茶匙
生粉 2 湯匙

450g shrimps
2–3 sticks spring onion (in shreds)
1/3 cup tomato paste
2 tbsps minced garlic
1 tbsp chopped chili
2 tsps minced ginger
2 tbsps caltrop starch

⦾ 醃料 | Marinade

糖 1/2 茶匙
鹽 1/4 茶匙
胡椒粉少許

1/2 tsp sugar
1/4 tsp salt
Dash of pepper

⦾ 調味料 | Seasonings

糖 1 湯匙
辣椒汁 2 茶匙
鹽 1/2 茶匙

1 tbsp sugar
2 tsps chili sauce
1/2 tsp salt

⦾ 芡汁 | Thickening

2 茶匙生粉
1/3 杯上湯

2 tsps caltrop starch
1/3 cup stock

⦾ 做法 | Method

1. 海中蝦去殼，去腸，留尾，沖洗乾淨，用廚房紙抹乾水份。
2. 在蝦中間位置切開，把蝦尾穿過，下醃料拌勻，置雪櫃中冷凍 1 小時。
3. 取出海蝦，加入生粉拌勻。
4. 燒一鍋油，投入海蝦炸至挺身，取出瀝油。
5. 熱鑊下油 1 湯匙，爆香蒜茸、薑茸和辣椒碎，加入芡汁煮稠，下調味，倒入炸蝦快束兜勻，即可上碟。

1. Shell shrimps and remove intestines, keep tails, Wash and wipe dry with kitchen paper.
2. Slit at back. Keep the tail. Then put the tail through the slit. Marinate and mix well, keep in refrigerator for an hour.
3. Take out the shrimps, add caltrop starch and mix well.
4. Heat a wok of oil, dump shrimps in and deep-fry until hard, remove and drain.
5. Pour 1 tbsp of oil in wok, Stir-fry minced garlic, ginger and chopped chili, sprinkle caltrop starch and thicken. Season and dump shrimps in, stir-fry and serve.

金盞明蝦

Fried Shrimps in Golden Bowl

4~6 人
Serves 4~6

20~25 分鐘
20~25 minutes

材料 | Ingredients

天使麵 120 克	120g capelli d' angelo
大蝦 8 隻	8 large shrimps
蘆筍 8 條	8 asparaguses
燻肉 8 片	8 slices smoked meat
紅椒碎少許	Pinch of minced red chili
胡椒粉少許	Pinch of red chili powder
黑胡椒碎少許	Pinch of crushed black peppercorn
鹽少許	Pinch of salt

做法 | Method

1. 將天使麵煮至七成熟，瀝乾水分。
2. 用手將一小撮天使麵扭成籃形。
3. 放在火鍋杓中，用油炸至金黃色；用油炸香燻肉；把胡椒粉、鹽和大蝦拌勻。
4. 燒熱橄欖油，放入蒜茸炒香；並煎熟大蝦；蘆筍只要頂段，用熱水灼熟。
5. 在碟上順序放上燻肉、天使籃、大蝦和蘆筍，再灑上鹽、黑胡椒碎及紅椒碎調味。

1. Cook angel hair until medium done. Drain.
2. With hands twist a bundle of angel hair into a basket shape.
3. Deep-fry angel hair basket until golden. Deep-Fry smoked meat until aromatic. Marinate shrimps with black pepper and salt.
4. Heat some olive oil. Sauté garlic until aromatic. Shallow-fry shrimps until done. Remove bottoms parts of asparaguses and scald.
5. Place smoked meat, angel hair basket, shrimps and asparagus on a dish in sequence. Season with salt, black pepper and red chili powder.

4~6 人
Serves 4~6

30~35 分鐘
30~35 minutes

蘆筍炒蝦球

Stir-fried Shrimps and Asparagus

⬤⬤ 材料｜Ingredients

中蝦 450 克	450g fresh shrimps
鮮蘆筍 6 條	6 fresh asparagus
葱白 1 條	1 white scallion
薑 2 片	2 slices ginger
甘筍花數片	Some slices carrot
油 5 湯匙	5 tbsps oil
酒 1 茶匙	1 tsp wine

⬤⬤ 蝦調味料｜Shrimp Seasonings

酒、粟粉各 1 茶匙

1 tsp wine
1 tsp caltrop starch

⑩ 蘆筍調味料 | Asparagus Seasonings

鹽 1/3 茶匙
水 3 湯匙

1/3 tsp salt
3 tbsps water

⑩ 芡汁 | Thickening

麻油、粟粉各 1 茶匙
鹽、糖各 1/3 茶匙
水 2 湯匙

1 tsp sesame oil
1 tsp caltrop starch
1/3 tsp salt
1/3 tsp sugar
2 tbsps water

⑩ 做法 | Method

1. 蝦洗淨，用稀鹽水（2 杯水加 1/2 茶匙鹽）浸 20 分鐘；然後去頭剝殼，從背部用刀剖開一半，下調味料拌勻。

2. 蘆筍去硬皮，洗淨；葱白切長段。

3. 燒熱 2 湯匙油，下蘆筍略炒，加調味料，大火炒熟，取出，排放碟上。

4. 把 3 湯匙油燒熱，用中火把蝦仁炒熟，盛起。

5. 原鑊爆香薑、葱白，放下蝦仁及甘筍花，潷酒，炒勻，去掉薑及葱白，最後下芡料兜勻，放在鮮蘆筍上即成。

1. Wash shrimps, soak in thin salt water (1/2 tsp of salt and 2 cups of water) for 20 minutes. Shell, cut their backs without separation, add seasonings and mix well.

2. Remove hard skins from asparagus and clean; cut white scallion into lengths.

3. Bringing 2 tbsps oil to a boil and stir-fry asparagus. Season. Stir-fry over high heat until cooked. Arrange onto a plate.

4. Bring 3 tbsps of oil to a boil, stir-fry shrimps quickly over medium heat until done. Set aside.

5. Sauté white scallion and ginger, dump carrot and shrimps in, sprinkle wine in, discard ginger and scallion, add thickening and serve.

蒜茸辣椒炒蟹

Stir-fried Crab with Garlic and Chili

◯◯ 材料 | Ingredients

蟹 2 隻	2 crabs
鹹蛋黃 2 隻 （蒸熟，磨茸）	2 salted egg yolks, steamed and mashed
雞蛋 1 隻	1 egg
薑茸 1 湯匙	1 tbsp minced ginger
蒜茸 1 湯匙	1 tbsp minced garlic
乾葱茸 1 湯匙	1 tbsp minced dried shallot
辣椒茸 1 湯匙	1 tbsp minced red chili
辣椒醬 1 湯匙	1 tbsp chili sauce

4~6 人
Serves 4~6

20~25 分鐘
20~25 minutes

34

◯◯ 醃料 | Marinade

鮮露 2 茶匙

2 tsps maggie sauce

◯◯ 調味料 | Seasonings

茄汁 1 湯匙
糖 2 茶匙
鹽 1/2 茶匙

1 tbsp ketchup
2 tsps white sugar
1/2 tsp salt

◯◯ 芡汁 | Thickening

生粉 2 茶匙
上湯 1/3 杯

2 tsps caltrop starch
1/3 cup chicken broth

◯◯ 做法 | Method

1. 把蟹劏洗乾淨，切件，加醃料醃 5 分鐘。

2. 燒熱鑊，放 5-6 湯匙油，爆香所有香料，放入蟹件煎至出香味，
   潲酒，加入調味料兜勻，蓋鑊蓋焗 15 分鐘。

3. 放入芡汁料兜勻煮至濃稠，再焗 5 分鐘或至熟透，打入雞蛋液兜
   勻，即可上碟。

1. Wash crabs. Chop into chunks. Marinate for 5 minutes.

2. Heat 5-6 tbsps of oil in wok. Sauté spices until aromatic.
   Add crabs. Sauté until aromatic. Sprinkle with some
   cooking wine. Add seasonings. Sauté well. Cover. Cook for
   15 minutes.

3. Add thickenings. Stir well. Cook for a further 5 minutes or
   until done. Add beaten egg. Stir well. Dish.

金沙紅袍

Golden Crab

材料 | Ingredients

膏蟹 900 克	900g crab
鹹蛋黃 6 隻	6 salted egg yolks
薑 40 克	40g ginger
蒜頭 60 克	60g shallots
辣椒 1 隻	1 chili
葱 2–3 條	2–3 sticks spring onion
生粉 3 湯匙	3 tbsps caltrop starch

◯◯ 調味料 | Seasonings

糖 2 茶匙
鹽 1 茶匙

2 tsps sugar
1 tsp salt

◯◯ 做法 | Method

1. 薑、蒜頭和辣椒洗淨，剁茸；把葱洗淨，切碎。

2. 鹹蛋黃蒸熟，壓碎。

3. 膏蟹內外洗刷乾淨，去掉砂囊、內臟和肺部，瀝乾，用生粉拌勻。

4. 把油燒至 220℃，投入膏蟹，以大火炸至八成熟，取出瀝油。

5. 燒 3 湯匙油，放入薑茸、蒜茸和辣椒茸爆香，加入鹹蛋黃碎，立即倒入膏蟹和調味料，快速翻炒，撒上葱粒，上碟。

1. Wash ginger, shallots and chili and chop finely. Wash spring onion and cut into sections.

2. Steam salted egg yolks to well done and mash.

3. Brush and wash crab, remove the stomach, intestines and gills. Drain and mix well with caltrop starch.

4. Heat oil to 220℃, dump the crabs in and deep-fry to 80% done, remove and drain.

5. Pour 3 tbsps oil in wok, stir-fry minced ginger, garlic and chopped chili, add yolk paste, then immediately put crabs and seasonings in to stir well, sprinkle spring onion and dish up to serve.

雞油蒸醃仔蟹

Steamed Mud Crabs with Chicken Fat

4~6 人
Serves 4~6

5~10 分鐘
5~10 minutes

材料 | Ingredients

泥蟹 4-6 隻
蔥 2-3 棵
薑片 40 克
蒜頭 2 粒
雞膏（固體雞油）20 克

4-6 mud crabs
2-3 sprigs spring onion
40g sliced ginger
2 cloves garlic
20g chicken fat

調味料 | Seasonings

鹽少許
糖少許

Pinch of salt
Pinch of sugar

蘸汁料 | Dipping Sauce

大紅浙醋 1/3 杯

1/3 cup Zhejiang red vinegar

做法 | Method

1. 泥蟹劏洗乾淨，斬件，蟹箝略拍碎。
2. 蔥洗淨，切段；蒜頭切片。
3. 燒熱鑊，把雞膏煎成雞油，放進蒜片爆香，熄火。
4. 在蟹件放上薑片和蔥段，撒點鹽和糖，再淋上雞油。
5. 燒一鍋水，放上蟹，用大火蒸 15-20 分鐘，取出，與蘸汁享用即可。

1. Wash crabs. Chop. Crack the pincers.
2. Wash spring onion. Strip. Slice garlic cloves.
3. Melt chicken fat in wok. Add garlic slices. Sauté until aromatic. Remove from heat.
4. Place ginger and spring onion on top of crabs. Sprinkle with salt and sugar. Pour melted chicken fat on top.
5. Boil some water. Steam crabs over high heat for 15-20 minutes. Serve with some red vinegar.

釀蟹蓋

Stuffed Crab Shell

◯◯◯ 材料 | Ingredients

膏蟹 2 隻
免治豬肉 50 克
雞蛋黃 2 個

2 pcs mud crab
50g minced pork
2 egg yolks

醃料 | Marinade

生抽 2 茶匙
油 2 茶匙
生粉 1/2 茶匙
糖少許
胡椒粉少許
水 1/2 茶匙

2 tsps light soy sauce
2 tsps oil
1/2 tsp caltrop starch
Pinch of sugar
Pinch of pepper
1/2 tsp water

調味料 | Seasonings

胡椒粉少許
鹽少許

Pinch of pepper
Pinch of salt

做法 | Method

1. 膏蟹洗淨，去鰓，蟹膏取出留用。
2. 蟹身和蟹蓋隔水蒸 10 分鐘，取出待涼後拆肉。
3. 免治豬肉醃 15 分鐘。
4. 燒 1 湯匙油熱，下免治豬肉爆香，取出。
5. 免治豬肉、蟹肉、蛋黃和調味拌勻，釀入蟹蓋中，以大火蒸 8 分鐘。

1. Wash crab and remove the gill. Take the fat out and set aside.
2. Steam crab for 10 minutes, leave to cool and retain the flesh.
3. Marinate minced pork for 15 minutes.
4. Heat 1 tbsp of oil, stir-fry minced pork until aromatic and remove.
5. Mix minced pork, crab meat, egg yolk and seasonings well and stuff in the crab shell. Steam over high heat for 8 minutes and serve hot.

香蔥蟹肉脆鍋巴

Crab with Crispy Puff Rice

4~6 人
Serves 4~6

25~30 分鐘
25~30 minutes

⃝⃝⃝ 材料 | Ingredients

花蟹 1–2 隻
即食鍋巴 10 件
薑 4 大片
葱 2 棵
雞蛋 1 隻

1–2 pcs spotted crabs
10 slices rice cake
4 big slices ginger
2 sprigs spring onion
1 egg

⃝⃝⃝ 芡汁 | Thickening

生粉 1 茶匙
生抽 1 茶匙
鹽 1/2 茶匙
糖 1/2 茶匙
胡椒粉少許
清水 2 湯匙

1 tsp caltrop starch
1 tsp light soy sauce
1/2 tsp salt
1/2 tsp sugar
Pinch of pepper
2 tbsps water

⃝⃝⃝ 做法 | Method

1. 蟹劏好洗淨，蒸 10–15 分鐘至熟，待涼後拆肉備用。
2. 薑切片；葱切段；雞蛋打散。
3. 燒熱 2 湯匙油，爆香薑和葱，撈取薑片。
4. 下蟹肉和芡汁煮滾，加入蛋液兜勻，即可放在鍋巴上桌享用。

1. Wash crab and steam for 10–15 minutes until cooked. Leave to cool, retain the flesh and set aside.
2. Cut ginger into slices, spring onion into sections and beat egg.
3. Heat 2 tbsps of oil, sauté ginger and spring onion until aromatic and remove sliced ginger.
4. Add crab meat and thickening and bring to the boil. Toss with beaten egg until well combined, then place crabs on top of rice cake and serve.

4~6 人
Serves 4~6

25~30 分鐘
25~30 minutes

古法蒸蟹缽

Steamed Crab Casserole in
Traditional Style

材料 | Ingredients

花蟹 / 泥蟹 2 隻
免治豬肉 250 克
雞蛋 2 隻
蒜茸 2 湯匙
蔥粒 1 湯匙

2 spotted crabs or mud crabs
250g minced pork
2 eggs
2 tbsps minced garlic
1 tbsp chopped spring onion

調味料 | Seasonings

生粉 2 茶匙
油 2 茶匙
酒 2 茶匙
糖 1 茶匙
鹽 1/2 茶匙
胡椒粉適量

2 tsps caltrop starch
2 tsps cooking oil
2 tsps cooking wine
1 tsp white sugar
1/2 tsp salt
Pinch of pepper

做法 | Method

1. 蟹劏洗乾淨，斬件，蟹箝略拍碎。

2. 豬肉和蟹膏拌勻，加入調味料和蒜茸拌勻。再打入雞蛋拌勻，並攪至起膠，放入蟹件於肉碎內。

3. 燒一鍋滾水，放入蟹缽以大火蒸 15–20 分鐘。再把蟹缽直接放火爐上以慢火燒至傳出香味約 5–10 分鐘，撒上蔥粒裝飾。

1. Wash crabs. Chop. Crack the pincers.

2. Combine the crab's fat and ground pork. Stir well. Add seasonings and minced garlic. Stir well. Add beaten eggs. Whisk until it forms sticky paste. Add crabs. Stir well.

3. Boil water. Steam the crabs in a casserole over high heat for 15–20 minutes. Then heat the casserole directly over low heat until aromatic, about 5–10 minutes. Sprinkle with chopped spring onion.

鮮蝦油豆腐

Braised Shrimps with Yaki Aburaage

材料 | Ingredients

鮮蝦 4–6 隻
罐裝筍條 20 克
蘆筍 4–6 條
日本油揚 / 豆泡 4–6 個
韭菜花 4–6 條

4–6 shrimps
20g canned young bamboo shoots
4–6 asparaguses
4–6 yaki aburaage/beancurd puffs
4–6 sprigs chive

4~6 人
Serves 4~6

15 分鐘
15 minutes

⟨⟨⟨ 煨油揚 / 竹筍料 | For Braising

雞粉 1/2 茶匙
清水 1 杯

1/2 tsp chicken powder
1 cup water

⟨⟨⟨ 芡汁 | Thickening

清湯 1/4 杯
糖 1/4 茶匙
鹽 1/4 茶匙
生粉 1 湯匙

1/4 cup chicken broth
1/4 tsp white sugar
1/4 tsp salt
1 tbsp caltrop starch

⟨⟨⟨ 做法 | Method

1. 鮮蝦去腸，洗淨，上籠以大火蒸 5-8 分鐘至蒸熟，去殼。

2. 蘆筍洗淨；用熱水把油揚的油份沖走；把韭菜花焯軟。

3. 煨料煮滾，把蘆筍、筍條和油揚分別煮片刻，取出過冷，壓出水份備用。

4. 把一條蘆筍、筍條和一隻蝦放在油揚上，以韭菜花綁在油揚上。放進蒸籠以大火蒸 5 分鐘，取出，倒去水份。把芡汁料調勻煮至濃稠，淋在油揚上即成。

1. Remove intestines from shrimps. Wash. Steam over high heat for 5–8 minutes or until done. Unshell shrimps.

2. Wash asparaguses. Rinse yaki aburaage with hot water. Soak chive in hot water until soft.

3. Bring the ingredients for braising to a boil. Add asparaguses, bamboo shoots and yaki aburaage. Cook for a while. Remove from heat. Rinse in cold water. Squeeze excess water. Set aside.

4. Wrap one piece of asparagus, bamboo shoot and shrimp with one yaki aburaage. Tie up with the blanched chive. Steam over high heat for 5 minutes or until done. Drain excess water. Combine the thickening. Stir well. Cook until the sauce begins to thicken. Pour onto the food. Serve.

雜錦海鮮煲

Seafood Casserole

〇〇 **材料 | Ingredients**

蝦 2 隻	2 shrimps
帶子 2 隻	2 scallops
日本魚蛋 3 片	3 slices Japanese fish cake
蟹柳 2 條	2 sticks crab
鮮魷 1 隻	1 squid
素魚翅適量	Pinch of Vegetarian fish fin
蝦乾 10 隻	10 dried shrimps
瑤柱 2 粒	2 dried scallops
白菜 2 棵	2 sprigs white cabbage
小椰菜 2 個	2 brussels sprout
蘆筍 2 棵	2 sprigs asparagus
本菇 / 靈芝菇少許	Pinch of Shimeji mushroom
清水 2 杯	2 cups water
薑 2 片	2 slices ginger

調味料 | Seasonings

鹽適量
胡椒粉少許

Pinch of salt
Pinch of pepper powder

做法 | Method

1. 鮮魷洗淨，撕去外膜，先剥花再切片；瑤柱、蝦乾用水浸軟；其他材料洗淨。

2. 在砂鍋中燒熱 2 湯匙油，爆香薑片，下清水煮滾，放入瑤柱和蝦乾煮 15 分鐘，便成上湯。

3. 先將白菜、小椰菜、本菇放入上湯煮 2 分鐘，再放餘下材料煮滾，下調味料即可。

1. Wash squid, remove sheath, make cross cuts and slice. Swell dried scallops and dried shrimps. Wash other ingredients.

2. Heat 2 tbsps oil in a casserole. Sauté ginger slices until aromatic. Add water and bring to a boil. Add swollen scallops and shrimps and cook for 15 minutes to make broth.

3. Add Chinese cabbage, brussels sprouts and mushrooms and cook for 2 minutes. Then add other ingredients and bring to a boil. Add seasonings. Serve.

蟹肉燴素翅

Braised Imitation Shark-fin with Crab

4~6 人
Serves 4~6

15~20 分鐘
15~20 minutes

⦿ 材料 | Ingredients

蟹 1 隻
素翅 300 克
蘆筍粒 50 克
甘筍絲 50 克
雞蛋白 1 隻
上湯 1 公升

1 crab
300g imitation shark-fin
50g diced asparaguses
50g shredded carrots
1 egg white
1 litre chicken broth

⦿ 芡汁料 | Thickening

上湯 1/3 杯
生粉 3 湯匙
糖 1 茶匙
鹽 1/2 茶匙
胡椒粉適量

1/3 cup chicken broth
3 tbsps caltrop starch
1 tsp white sugar
1/2 tsp salt
Pinch of pepper

⦿ 做法 | Method

1. 把蟹劏洗乾淨，擺放碟上，放蒸籠以大火蒸 10-15 分鐘，取出拆肉。

2. 把蘆筍和甘筍洗淨，蘆筍切粒，甘筍切絲，汆水過冷，備用。

3. 把上湯與調味料同置鍋中燒滾，期間不斷攪拌至濃稠，加入蟹肉和素翅煮滾，再放入甘筍和蘆筍煮滾，熄火。

4. 倒入雞蛋白攪至全熟，盛起，伴以香醋和芫茜享用。

1. Wash crab. Chop into chunks. Steam crab over high heat for 10-15 minutes. Unshell the crab and obtain the flesh.

2. Wash asparaguses and carrot. Dice asparaguses and shred carrot. Blanch. Then rinse in cold water.

3. Bring chicken broth and seasonings to a boil. Keep stirring until thicken. Add crab meat and imitation shark-fin. Bring to a boil. Add carrot and asparaguses. Bring to a boil again. Remove from heat.

4. Add egg white and keep stirring until done. Serve with some vinegar and parsley.

金黃蟹箝

Deep-fried Crab Claws

材料 | Ingredients

蟹箝 12 隻	12 crab claws
中蝦 450 克	450g shrimp
肥豬肉 40 克	40g fat pork
馬蹄 40 克	40g water chestnuts
雞蛋 2 隻	2 eggs
麵粉 4 湯匙	4 tbsps flour
麵包糠 4 湯匙	4 tbsps breadcrumbs
鹽 1 茶匙	1 tsp salt

4~6 人
Serves 4~6

15~20 分鐘
15~20 minutes

◯◯◯ 調味料 | Seasonings

雞蛋白 1 隻	1 egg white
生粉 2 湯匙	2 tbsps caltrop starch
糖 1 茶匙	1 tsp sugar
胡椒粉 1/4 茶匙	1/4 tsp pepper
麻油少許	Some sesame oil

◯◯◯ 做法 | Method

1. 肥豬肉洗乾淨，切粒；馬蹄去皮，用刀拍爛。
2. 蟹箝洗淨，以大火蒸 10 分鐘，用鐵箝把殼略夾碎，但不要夾斷。
3. 海中蝦去殼，去腸，用少許生粉拌勻，沖水，用乾布抹乾水份，置雪櫃中冷藏 1 小時，備用。
4. 取出蝦肉，用刀拍爛，放入大碗中，加鹽，以順時針方向大力攪勻，直至蝦肉呈膠結狀。加入調味料、馬蹄肉和肥豬肉，攪勻。
5. 蝦膠裹在蟹箝上，成球狀，依次滾上麵粉、雞蛋和麵包糠。
6. 把油燒至 200℃熱，放入蟹箝炸至金黃約 8-10 分鐘，便可上碟。

1. Wash fat pork, cut into pellets. Peel water chestnuts and smash.
2. Wash crab claws, steam over high heat for 10 minutes, break the crab claws with tongs but keep each claw whole.
3. Shell shrimps and remove the intestine, mix well with a dash of caltrop starch. Rinse under water, wipe dry with kitchen cloth and freeze in a refrigerator for an hour.
4. Take out shrimps, smash and transfer into a big bowl, add salt, stir clockwise to form shrimp paste. Add seasonings, water chestnuts and fat pork to stir well.
5. Coat crab claws with shrimp paste to be balls, then coat with flour, egg and breadcrumbs in turn.
6. Heat oil at 200℃, dump crab claws in and deep-fry for 8-10 minutes until golden, dish and serve.

椒鹽蝦

Spicy Shrimps

⫯⫯⫯ 材料 | Ingredients

蝦 480 克
辣椒 2 隻
蒜茸 2 湯匙
葱粒 1 湯匙
薑茸 1 茶匙

480g shrimps
2 red chilies
2 tbsps minced garlic
1 tbsp chopped spring onion
1 tsp minced ginger

醃料 | Marinade

鮮露 2 茶匙
生粉 2 茶匙
鹽 1/2 茶匙

2 tsps Maggi sauce
2 tsps caltrop starch
1/2 tsp salt

調味料 | Seasonings

味椒鹽 1/2 茶匙
糖 1/2 茶匙

1/2 tsp chili salt and pepper
1/2 tsp white sugar

做法 | Method

1. 蝦在背部挑去蝦腸，洗淨抹乾，加入鹽和鮮露拌勻，醃 10 分鐘。

2. 辣椒去籽，洗淨，切碎。

3. 鑊燒熱，下 3-4 湯匙；把蝦用生粉 2 茶匙拌勻，放在鑊中以中火煎至兩面金黃，取出。

4. 原鑊再下 2 湯匙油燒熱，下辣椒茸、蒜茸和薑茸爆香，再放入已煎熟的蝦，下調料兜勻，撒下葱粒。

1. Remove the intestines from shrimps. Wash. Blot dry. Marinate with the salt and Maggie sauce for 10 minutes.

2. Seed the red chilies. Rinse and chop.

3. Heat 3-4 tbsps of oil in wok. Coat shrimps with 2 tsps of caltrop starch. Deep-fry shrimps over moderate heat until golden brown.

4. Add 2 tbsps of oil to the same wok. Sauté chopped red chili, minced garlic and minced ginger until aromatic. Add shrimps and seasonings. Sauté well. Sprinkle with some chopped spring onion.

鮮蝦墨魚卷

Crisp Shrimp and Cuttlefish Rolls

材料 | Ingredients

墨魚膠 450 克
蒸熟蝦粒 80 克
鮮腐皮 1 張
急凍雜菜粒 1 湯匙（汆水）

450g cuttlefish paste
80g steamed shrimps dices
1 fresh bean curd sheet
1 tbsp blanch frozen mixed vegetables

4~6 人
Serves 4~6

20~25 分鐘
20~25 minutes

調味料 | Seasonings

雞蛋白 1 隻
生粉 1 茶匙
鹽 1/2 茶匙
糖 1/2 茶匙
胡椒粉適量

1 egg white
1 tsp caltrop starch
1/2 tsp salt
1/2 tsp white sugar
Pinch of pepper

雜果沙律料 | Mixed Fruit Salad Dressing

芒果粒 150 克
蘋果粒 150 克
蛋黃醬適量
煉奶少許

150g diced mango
150g diced apple
Some Mayonnaise
Some concentrated milk

做法 | Method

1. 鮮腐皮用濕布抹淨，剪成小塊。

2. 把墨魚膠加入調味料，放入雜菜粒，拌勻置冰箱中冷藏 30 分鐘。

3. 手濕點油或水，將墨魚膠抹在腐皮上，放上蝦粒，捲成春卷狀。

4. 油燒至 8 成熱，放墨魚卷炸至金黃，取出瀝油，上碟。把沙律料拌勻，與墨魚卷同吃。

1. Rub bean curd sheet with a wet cloth. Cut into small pieces.
2. Combine seasonings, cuttlefish paste and mixed vegetables. Stir well. Refrigerate for 30 minutes.
3. Wet hands with oil or water. Stuff cuttlefish paste onto beancurd sheet. Place some diced shrimps on top. Roll up.
4. Heat oil until 80% hot. Deep-fry spring rolls until golden. Drain excess oil. Dish. Combine ingredients of salad dressing. Serve.

蒜香燒大蝦

Grilled Prawns with Garlic

材料 | Ingredients

大蝦 4-6 隻
巴士馬芝士粉適量
日式麵包糠適量

4-6 prawns
Some parmesan cheese powder
Some Japanese breadcrumbs

醃料 | Marinade

雞蛋白 1 隻
砂糖 1 茶匙
味椒鹽 1 茶匙
蒜鹽 1/2 茶匙
胡椒碎適量

1 pc white egg
1 tsp castor sugar
1 tsp spicy salt
1/2 tsp garlic salt
Some pepper

蘸汁料 | Dipping Sauce

青檸 1 個（切角）
日式芝麻醬適量

1 lime (cut in wedges)
Japanese sesame sauce

入廚貼士 | Cooking Tips

- 用醃料將大蝦醃透，煎至兩面轉色，才塗點雞蛋白，再分別放上麵包糠和芝士，轉到焗爐中燒烤時，能保持大蝦的肉汁。

- Marinade prawns and seal until turned colour, brush a bit of egg white, bread crumb and shredded cheese accordingly. Transfer to an oven for baking that will have a better cooking result to keep juice of the prawns.

做法 | Method

1. 大蝦去頭和腸，並在蝦背上劏開，沖洗乾淨，抹乾。
2. 加入醃料拌勻，醃 5 分鐘，穿上竹籤，放上麵包糠，撒上芝士粉。
3. 焗爐預熱至 200℃，放入蝦串焗 10-15 分鐘或至金黃色，伴以蘸汁料享用。

1. Remove the heads and intestines of the prawns. Slit on the back of the prawns, wash and pat dry.
2. Mix with marinade and leave it for 5 minutes. Thread the prawns onto bamboo skewer, coat with breadcrumbs and sprinkle with cheese powder.
3. Preheat the oven to 200℃. Place prawn skewers into the oven and bake for 10 to 15 minutes until golden. Serve with dipping sauce.

芝士焗龍蝦尾

Baked Lobster Tail with Cheese

4~6 人
Serves 4~6

20~25 分鐘
20~25 minutes

材料 | Ingredients

龍蝦尾 2-3 條	2-3 lobster tails
忌廉雞湯 1 杯	1 cup canned creamy chicken soup
蒙沙娜芝士 50 克	50g Mozzarella cheese
蘑菇片 50 克	50g sliced mushrooms
洋蔥絲 50 克	50g shredded onion
青椒絲 50 克	50g shredded green bell pepper
蒜茸 2 茶匙	2 tsps minced garlic
清水 1/4 杯	1/4 cup water

調味料 | Seasonings

番茜碎 2 茶匙	2 tsps chopped fresh parsley
糖 1 茶匙	1 tsp white sugar
黑胡椒碎適量	Pinch of black pepper

做法 | Method

1. 龍蝦尾解凍，洗淨，抹乾，用少許胡椒和鹽調味，以大火煎至兩面轉色，取出。

2. 燒鍋下油 1-2 湯匙，爆香洋蔥、蒜茸和青椒絲，下蘑菇片略炒，加入忌廉雞湯和清水煮至濃稠，下調味料煮滾。

3. 把龍蝦尾放焗盤上，加入忌廉雞湯，撒上芝士，放入已預熱的焗爐至 200℃焗 10-15 分鐘，即成。

1. Defrost lobster tails. Wash. Blot dry. Season with some pepper and salt. Shallow-fry over high heat until color changed. Set aside.

2. Heat wok with 1-2 tbsps oil. Sauté onion, garlic and green bell pepper until aromatic. Add mushrooms. Sauté for a while. Add creamy chicken soup and water. Cook until thicken. Add seasonings. Bring to a boil.

3. Put lobster tails on a baking pan. Add creamy chicken sauce. Sprinkle with some cheese. Preheat the oven to 200℃. Bake for 10-15 minutes. Serve.

鹽焗蟹

Baked Crabs with Salt

⬭⬭ 材料 | Ingredients

花蟹 / 泥蟹 4-6 隻
蔥 2-3 棵
薑片 40 克
粗鹽 600 克

4-6 spotted crabs or mud crabs
2-3 sprigs spring onion
40g sliced ginger
600g coarse salt

⊙⊙⊙ 蘸汁料 | Dipping Sauce

大紅浙醋 1/3 杯
紅糖 2 湯匙
薑茸 1 茶匙

1/3 cup Zhejiang red vinegar
2 tbsps cane sugar
1 tsp minced ginger

⊙⊙⊙ 做法 | Method

1. 泥蟹劏洗乾淨，斬件，蟹箝略拍碎。
2. 葱洗淨，切段。
3. 用薑片和葱段放在蟹上，再砌回蟹形，用焗餅紙包好蟹，放在墊有粗鹽的焗餅紙，再蓋上一層粗鹽，用錫紙包好。
4. 焗爐預熱至 200℃，放入蟹焗 20–25 分鐘，伴以已調勻的蘸汁料享用。

1. Wash mud crabs. Chop. Crack the pincers.
2. Wash spring onion. Strip.
3. Resemble the crabs. Place ginger and spring onion on top. Wrap each crab with a piece of baking paper layered with coarse salt. Then wrap it with a piece of metal foil layered with coarse salt.
4. Preheat an oven to 200°C. Bake the crabs for 20–25 minutes. Serve with the dip.

菠蘿咕嚕大蝦

Prawns in Sweet and Sour Sauce

〇〇〇 材料 | Ingredients

大蝦 4 隻
青、紅椒各 1/2 個
罐頭菠蘿 4 片
葱 1 條

4 pcs king shrimps
1/2 each green and red pepper
4 slices canned pineapple
1 spring onion

4~6 人
Serves 4~6

20~25 分鐘
20~25 minutes

炸漿料 | Deep-Fry Batter

雞蛋黃 1 隻
粟粉 1/2 杯

1 egg yolk
1/2 cup caltrop starch

芡汁 | Thickening

菠蘿汁 1/4 杯
茄汁 2 湯匙
糖 1 湯匙
生粉 1 茶匙

1/4 cup pineapple sauce
2 tbsps ketchup
1 tbsp sugar
1 tsp caltrop starch

做法 | Method

1. 大蝦修剪好，挑腸，沖淨，瀝乾。
2. 青、紅椒切角。菠蘿切片；葱切段。
3. 大蝦用雞蛋黃拌勻，逐一撲上粟粉，輕輕放入滾油中，炸至金黃色，取出瀝油。
4. 原鍋倒出炸油，讓鍋中剩油 2 湯匙，爆香葱段，加青紅椒及菠蘿，埋芡汁，待芡汁開始收乾時，放大蝦回鍋兜勻，即可上碟。

1. Trim prawns, devein, wash and pat dry.
2. Cut green and red pepper into wedges. Cut pineapples into slices and spring onion into sections.
3. Mix prawns with egg yolk thoroughly and coat with caltrop starch. Deep-fry the prawns in hot oil until golden and drain excess oil.
4. Pour oil out with 2 tbsps of oil left in the wok. Sauté spring onion, add green and red peppers and pineapples, then stir in sauce. When the sauce starts to dry up, toss with the prawns and serve.

入廚貼士 | Cooking Tips

• 如果用新鮮菠蘿入饌，可以挖通菠蘿，做成菠蘿船，增加食趣。

• If fresh pineapple is being used, you may scoop out the flesh and make a boat-like container for this dish.

Deep-fried Shrimp Balls

錦繡蝦球

材料 | Ingredients

中蝦 6 隻
蝦肉 240 克
方包 4 片
雞蛋 1 隻
蟹子適量

6 pcs medium shrimps
240g peeled shrimp
4 slices bread
1 egg
Some crab roe

醃料 | Marinade

生粉 2 茶匙
鹽 1 茶匙
胡椒粉 1/2 茶匙

2 tsps caltrop starch
1 tsp salt
1/2 tsp pepper

做法 | Method

1. 將蝦肉挑腸，洗淨，抹乾水份，剁碎，放入碗內，加入醃料攪勻，順時鐘方向攪至起膠，放入冰箱內冷藏 30 分鐘。

2. 中蝦去殼，挑腸，洗淨，抹乾水份。

3. 方包切粒；雞蛋打散。

4. 取出蝦膠，包裹着中蝦，沾上蛋液，黏上麵包粒。

5. 燒油半鍋，放蝦多士炸至金黃色，撈起，放上少許蟹子即可。

1. Devein peeled shrimp, wash and wipe dry. Mash shrimp, place them in a bowl, add marinade in and mix well. Stir mashed shrimp in clockwise direction to make a sticky paste and chill for 30 minutes.

2. Peel and devein shrimps, wash and wipe dry.

3. Cut bread into cubes and beat eggs.

4. Take shrimp paste out, wrap a shrimp in, dip some beaten egg and then coat with bread cubes.

5. Heat half wok of oil, deep-fry shrimp toast until golden and dish up. Place some crab roe on and serve hot.

香蒜瀨尿蝦

Garlic Mantis Shrimp

4~6 人
Serves 4~6

15~20 分鐘
15~20 minutes

材料 | Ingredients

大瀨尿蝦 4 隻
4 pcs mantis shrimp

炸漿料 |
Deep-fry Batter

生粉 1/2 杯
1/2 cup caltrop starch

調味料 | Seasonings

蒜頭 2 個
指天椒 3 隻
鹽 2 茶匙

2 cloves garlic
3 pcs cayenne pepper
2 tsps salt

做法 | Method

1. 瀨尿蝦洗淨，然後燒一鑊水煮熟瀨尿蝦，抹乾。
2. 蒜頭去衣，剁成茸。
3. 燒熱油半鑊，將瀨尿蝦撲上生粉，下鑊炸至金黃色，盛起瀝乾。
4. 燒熱 3 湯匙油，放蒜茸炸至脆身，下紅椒絲爆香。瀨尿蝦回鑊炒勻，灑上鹽炒勻，即可。

1. Wash mantis shrimps, boil in boiling water until cooked and wipe dry.
2. Peel and chop garlic.
3. Heat half wok of oil, coat shrimps with caltrop starch, deep-fry until golden, remove and drain.
4. Heat 3 tbsps of oil, deep-fry garlic until crispy and stir-fry shredded cayenne pepper until aromatic. Return shrimps into the wok, stir in salt, toss until well combined and serve.

4~6 人
Serves 4~6

15~20 分鐘
15~20 minutes

乾燒大蝦

Pan-fried Prawns

材料 | Ingredients

大蝦 4-6 隻
珍珠米適量
紅珍珠米適量
紅米適量
水適量

4-6 prawns
Some round rice
Some red round rice
Some red long rice
Water

⊙⊙ 醃料 | Marinade

鹽 1/2 茶匙
黑胡椒碎適量
1/2 tsp salt
Some ground black pepper

⊙⊙ 醬汁料 | Sauce

泰式雞醬汁 2 湯匙
魚露 1 茶匙
青檸汁 1 茶匙

2 tbsps Thai Chicken Sauce
1 tsp fish sauce
1 tsp lime juice

⊙⊙ 做法 | Method

1. 珍珠米淘洗乾淨，加入紅珍珠米、紅米與清水同置煲中煮熟。
2. 大蝦去頭、去腸，洗淨，開邊，加醃料醃片刻，撲上少許生粉。
3. 熱鑊下油，放大蝦煎至金黃熟透，盛起。
4. 雞醬汁拌勻，與大蝦和紅米飯一齊享用。

1. Wash round rice thoroughly. Add red round rice, red long rice and water, and then cook them together in a rice cooker until well cooked.
2. Remove the heads and intestines of prawns, wash thoroughly, cut into halves, mix with marinade and marinate for a while. Then coat with some caltrop starch.
3. Put some oil into a heated frying pan. Pan-fry prawns until golden and cooked. Place on the dish.
4. Mix sauce ingredients well and then serve with the prawns and rice.

天婦羅蝦

Shrimp Tempura

⦿ 材料｜Ingredients

大蝦 4–6 隻
薄荷葉 1 棵
生粉適量

4–6 prawns
1 sprig peppermint
Pinch of caltrop starch

⦿ 天婦羅漿料｜For Tempura Batter

天婦羅粉 200 克
雞蛋黃 1 隻
冰水 700 毫升

200g tempura batter mix
1 egg yolk
700ml iced water

4~6 人
Serves 4~6

25~30 分鐘
25~30 minutes

◯◯ 醃料 | Marinade

鹽 1/2 茶匙
糖 1/2 茶匙
胡椒粉適量

1/2 tsp salt
1/2 tsp white sugar
Pinch of pepper

◯◯ 蘸汁料 | Dipping Sauce

天婦羅汁 100 毫升
白蘿蔔茸 1 湯匙

100ml tempura sauce
1 tbsp minced turnip

◯◯ 做法 | Method

1. 蝦洗淨、抹乾、挑腸、去殼留尾；頭和蝦身分開，用刀把蝦身略壓扁。
2. 蝦用醃料拌勻，醃 5–10 分鐘。
3. 天婦羅粉與冰水調勻，加入蛋黃拌勻，靜置 5–10 分鐘。
4. 燒油一鍋，把蝦沾少許生粉，再沾上天婦羅漿，放鑊炸透；蝦頭撲少許生粉，放油鑊炸好，取出瀝油。
5. 薄荷葉洗淨，沾上少許天婦羅漿，炸透，上碟與蘸料食用。

1. Shell prawns and remove intestines. Wash. Blot dry. Separate the heads and bodies. Slightly press the bodies.
2. Marinate the prawns for 5–10 minutes.
3. Combine the tempura batter mix and iced water. Stir well. Add egg yolk. Stir well. Set aside for 5–10 minutes.
4. Heat a pan of oil. Coat the prawns in sequence with the caltrop starch and tempura batter. Deep-fry until done. Dust some caltrop starch onto the prawn heads. Deep-fry until done.
5. Dip peppermint in tempura batter, deep-fry until done. Serve with the dip.

4~6 人
Serves 4~6

20~25 分鐘
20~25 minutes

麥皮蝦

Oatmeal Prawns

材料 | Ingredients

大蝦 600 克
即食麥皮 1/3 杯
雞蛋 1 隻（打散）
糖 1 茶匙
好立克粉 1 茶匙

600g prawns
1/3 cup instant oatmeal
1 egg, beaten
1 tsp white sugar
1 tsp Holick powder

醃料 | Marinade

鹽 1/2 茶匙
生粉適量

1/2 tsp salt
Pinch of caltrop starch

做法 | Method

1. 大蝦去腸，洗淨抹乾，在背部剅一刀，加入醃料醃 3-5 分鐘。

2. 把蝦用半煎炸法炸至金黃，取出瀝油，上碟。

3. 燒一鍋油，放進雞蛋，一邊煎一邊用筷子不斷轉動弄散，加入麥皮炸至金黃，取出瀝油。

4. 把麥片加糖和好立克粉拌勻，放在大蝦上即成。

1. Remove intestines from prawns. Wash. Blot dry. Make butterfly cuts at the back. Marinate for 3–5 minutes.

2. Shallow-fry the prawns until golden and dry. Drain well. Dish.

3. Heat a pan of oil. Add beaten egg. Keep stirring with chopsticks. Add the oatmeal. Deep-fry until golden. Drain well.

4. Combine oatmeal, sugar and Holick powder. Stir well. Place on top of the shrimps.

油炸軟殼蟹

Deep-fried Soft Shell Crab

材料 | Ingredients

軟殼蟹 4–6 隻
焯熟時菜適量
4–6 碗飯
脆炸乾葱茸適量
香蒜茸適量
生粉適量

4–6 soft shell crabs
Some blanched vegetables
4–6 bowls cooked rice
Some deep–fried minced shallots
Some deep–fried minced garlic
Some caltrop starch

4~6 人
Serves 4~6

20 分鐘
20 minutes

⬤⬤ 脆漿料 | Batter

炸粉 1/3 杯
冰水 2/3 杯
油 2-3 湯匙（後下）

1/3 cup batter powder
2/3 cup icy water
2-3 tbsps oil
(to be added in final stage)

⬤⬤ 調味料 | Seasonings

味椒鹽適量

Some spicy pepper

⬤⬤ 蘸汁料 | Dipping Sauce

沙律醬適量

Some mayonnaise

⬤⬤ 做法 | Method

1. 軟殼蟹解凍，去腮，洗淨，切件，用廚用紙抹乾，加入調味料醃 5-10 分鐘。

2. 炸粉與冰水調勻成粉漿，再加入油拌勻，靜置 10-15 分鐘，便成 脆漿。

3. 把軟殼蟹撲上生粉，沾上脆漿。燒油一鍋至八成滾，放入蟹件炸 透，取出瀝油，吃時灑一點味椒鹽。

4. 將軟殼蟹和白焯時菜同放碟中，可配拌了脆乾葱茸和蒜茸的白飯 供食，也可伴以沙律醬。

1. Defrost soft shell crabs, remove gills, wash thoroughly, cut into pieces, and then pat dry. Add seasonings and marinate for 5 to 10 minutes.

2. Mix batter powder with water thoroughly. Add in oil and blend well. Leave it for 10 to 15 minutes.

3. Coat crabs with caltrop starch and then dip the batter. Heat up a frying pan of oil until it becomes fairly hot. Deep-fry crabs until well cooked. Take out and drain excess oil. Serve with a sprinkle of spicy salt.

4. Place crabs, blanched vegetables on a plate. Serve with rice mixed with fried shallot and minced garlic. Or serve with mayonnaise.

蟹粥

Crab Congee

4~6 人
Serves 4~6

2 小時
2 hours

材料 | Ingredients

蟹 2 隻
白米 1 米杯
珍珠米 1 米杯
清水 4 公升

2 crabs
1 cup white rice
1 cup round rice
4 litres water

調味料 | Seasonings

鹽少許
胡椒粉少許

Some salt
Some pepper

入廚貼士 | Cooking Tips

• 用水蟹煲粥，味道更鮮甜。

• The congee is more delicious by cooking with "water crabs".

做法 | Method

1. 去除蟹的內臟，洗淨，切件。
2. 白米和珍珠米用淘洗乾淨，用適量水和 1 湯匙油浸 15–20 分鐘，瀝乾水份。
3. 清水加 3–4 隻瓦匙一同煲滾，加入白米煲 1 1/2 小時，放蟹件滾 15 分鐘，下調味料，即成。

1. Remove internal organs of the crabs, wash thoroughly and then cut into pieces.
2. Wash the rice and round rice thoroughly. Add some water and 1 tbsp of oil, soak for 15 to 20 minutes, and then drain excess water.
3. Place 3 or 4 china spoons and water in a pot and bring it to a boil. Add the rice and boil for 1 1/2 hours. Add in the crabs and boil for 15 minutes. Add seasonings and serve.

奶油龍蝦烏冬

Lobster Udon in Cream Sauce

材料 | Ingredients

龍蝦尾 4–6 隻　　　　上湯 250 毫升
忌廉雞湯 250 毫升　　烏冬 2 個
焯熟泰國蘆筍 100 克　麵粉 5 克
紅胡椒粒 10 克　　　　牛油 5 克
蒙沙華娜芝士碎 50 克　水 100 毫升

4-6 lobster tails
250ml cream of chicken
100g blanched Thai asparagus
10g red chilies dice
50g shredded mozzarella cheese
250ml stock
2 packets udon
5g flour
5g butter
100ml water

◯◯◯ 調味料 | Seasonings

胡椒碎 1 茶匙	1 tsp ground pepper
鹽 1/2 茶匙	1/2 tsp salt
糖 1/2 茶匙	1/2 tsp sugar
蒜鹽 1/2 茶匙	1/2 tsp garlic salt

◯◯◯ 做法 | Method

1. 烏冬用上湯焯熱，備用。
2. 熱鑊下牛油，炒香麵粉，徐徐加入清水煮成白汁，加入忌廉湯煮成濃汁。
3. 龍蝦尾解凍開邊，加入調味醃 5 分鐘，下點橄欖油煎至 4-5 成熟。
4. 烏冬放碟上，放上龍蝦尾，淋上忌廉汁，撒上芝士和紅胡椒粒，置已預熱至 190℃的焗爐中焗 10-15 分鐘或至龍蝦全熟，取出伴以蘆筍即成。

1. Blanch undo in hot stock until hot. Set aside.
2. Put butter into a heated frying-pan and then stir-fry the flour in it until it becomes aromatic. Add water gradually and stir until it turns into a white sauce. Add cream of chicken and cook until it becomes a thick sauce.
3. Defrost lobster tails, cut into halves, marinate with seasonings for 5 minutes, then pan-fry with some olive oil for 4 to 5 minutes.
4. Put udon on a plate, place lobster tails on top, pour creamy sauce over and then sprinkle with shredded cheese and red chili dice. Bake in a pre-heated oven at 190℃ for 10 to 15 minutes until lobsters are well cooked. Garnish with some asparagus. Ready to serve.

鮮蝦米線

Prawns Rice Vermicelli

材料 | Ingredients

蝦頭 500 克
大蝦 300 克
蝦殼 250 克
小魚乾 200 克
蟹柳 100 克
魷魚 150 克
米線 400 克

銀芽 100 克
上湯 3 公升
大蔥 100 克
芫茜頭 100 克
胡椒粒 50 克
蒜頭 10 粒

500g shrimp heads
300g prawns
250g shrimp shells
200g dried small fishes
100g crab sticks
150g squids
400g rice vermicelli

100g bean sprouts
3 litres broth
100g leeks
100g coriander heads
50g ground pepper
10 cloves garlic

4~6 人
Serves 4~6

30~45 分鐘
30~45 minutes

⦾ 調味料 | Seasonings

芫荽碎少許	Some chopped coriander
鹽少許	Some salt
胡椒粉少許	Some pepper

入廚貼士 | Cooking Tips

- 蝦湯用不完，可放入冰箱凍成冰粒，留待日後應用。
- The unused shrimp broth can be frozen in the freezer for later use.

⦾ 做法 | Method

1. 蝦頭和蝦殼洗淨，放入已預熱至 200℃的焗爐內焗 15–20 分鐘至乾身及有香氣排出。
2. 上湯放入蝦頭、蝦殼、胡椒粒、大蔥、芫荽頭和蒜頭煮滾，改中火煲 30–45 分鐘。
3. 把米線放入蝦湯煮熱；把蟹柳、大蝦和魷魚分別放上湯焯熟，把各已弄熱的材料同置碗中，淋上蝦湯和調味料，放上銀芽便可。

1. Wash shrimp heads and shells, and then bake in a pre-heated oven at 200℃ for 15 to 20 minutes until dried and aromatic.
2. Put shrimp heads, shells, pepper, leek, coriander heads and garlic into the broth and bring it to a boil. Turn the heat to medium and cook for 30 to 45 minutes.
3. Put rice vermicelli into shrimp broth and cook until it becomes hot. Blanch crab sticks, prawns and squids separately until well cooked. Put all the cooked Ingredients into a bowl, pour some shrimp broth and seasonings over, and then serve with bean sprouts.

燒大蝦撈麵

Fried Prawns with Noodles

⬤⬤⬤ 材料 | Ingredients

大蝦 4-6 隻　　　　　蛋麵 2 個
烘香中蝦頭 450 克　　上湯 2 杯
蝦膏 20 克　　　　　薑絲 20 克
烘香大地魚 20 克　　　葱絲 20 克
胡椒粒 5 克

4-6 prawns
450g medium shrimp heads (roasted)
20g shrimp paste
20g dried plaice (roasted)
5g ground pepper
2 egg noodles
2 cups broth
20g ginger (shredded)
20g spring onion (shredded)

調味料 | Seasonings

胡椒碎 1 茶匙
鹽 1/2 茶匙
糖 1/2 茶匙
蒜鹽 1/2 茶匙

1 tsp ground pepper
1/2 tsp salt
1/2 tsp sugar
1/2 tsp garlic salt

撈麵料 | Noodle Sauce Ingredients

蠔油 3 湯匙
熟油 1 湯匙

3 tbsps oyster sauce
1 tbsp cooked oil

入廚貼士 | Cooking Tips

- 蝦膏在售賣急凍海鮮的店舖有售。
- Shrimp paste is available for sale in frozen seafood stores.

做法 | Method

1. 把中蝦頭、蝦膏、大地魚、胡椒粒和上湯同置煲中滾 20 分鐘。
2. 蛋麵放滾水焯 2–4 分鐘,取出過冷,加點熟油拌勻,用上湯淋熱。
3. 大蝦開邊,加調味料醃 5 分鐘,用鑊下油煎熟。
4. 把麵、薑、葱和大蝦同置碟上,加入蠔油上桌,湯另上伴吃。

1. Boil medium shrimp heads, shrimp paste, dried plaice, ground pepper and broth in a pot for 20 minutes.
2. Blanch egg noodles in boiling water for 2 to 4 minutes, take out, rinse, and then mix it with some cooked oil. Pour some hot broth on top.
3. Cut prawns into halves, add seasoning and leave it for 5 minutes. Pan–fry prawns until well cooked.
4. Place the noodles, ginger, spring onion and prawns in a dish, add some oyster sauce, serve with some broth as accompaniment.

喇沙

Laksa

⦿⦿⦿ 材料 | Ingredients

蝦頭 500 克
蝦殼 250 克
小魚乾 200 克
熟蛋麵 400 克
煮熟豆腐泡 150 克
銀芽 100 克
上湯 3 公升
椰汁 250 克

大葱 100 克
芫荽頭 100 克
胡椒粒 50 克
紅咖喱醬 30 克
蝦膏 10 克
辣椒碎 10 克
蒜頭 10 粒
蒜茸 1 湯匙

500g shrimp heads
250g shrimp shells
200g dried little fishes
400g cooked egg noodles
150g beancurd puffs (cooked)
100g bean sprouts
3 litres broth
250g coconut milk

100g leek
100g coriander root
50g ground pepper
30g red curry paste
10g shrimp paste
10g chopped chili
10 cloves garlic
1 tbsp minced garlic

4~6 人
Serves 4~6

30 分鐘
30 minutes

調味料 | Seasonings

胡椒粉、芫荽碎各少許
Some pepper
Some chopped coriander

入廚貼士 | Cooking Tips

- 蝦湯熬成後，待吃時才加入椰汁煮湯，便不容易變壞。

- To prevent the broth from rotting, cook coconut milk with the shrimp broth shortly before serving.

做法 | Method

1. 蝦頭和蝦殼放入 200℃ 的焗爐內焗 15–20 分鐘，至乾身及有香氣。

2. 上湯放入蝦頭、蝦殼、胡椒粒、大葱、芫荽頭和蒜頭煮滾，改中火煲 30–45 分鐘。

3. 熱鑊下油 1 湯匙爆香蒜茸、紅咖喱醬、蝦膏和辣椒碎，注入蝦湯，加入椰汁煮滾，放入麵、豆腐泡和銀芽享用。

1. Bake shrimp heads and shells in oven at 200°C for 15 to 20 minutes until dry and aromatic.

2. Add shrimp heads, shrimp shells, ground pepper, leek, coriander root and garlic into the broth and bring it to a boil. Lower the heat to medium and boil for 30 to 45 minutes.

3. Heat 1 tablespoonful oil in frying-pan. Add minced garlic, red curry paste, shrimp paste and chopped chili, and sauté until aromatic. Add shrimp broth and coconut milk and bring it to a boil. Place noodles, beancurd puffs and bean sprouts. Serve.

蝦蟹滋味

編著
飲食同盟會

編輯
紫彤

美術設計
Ceci

排版
劉葉青

翻譯
Rosanna　Tracy

攝影
Wilson Wong

出版者
萬里機構出版有限公司
香港鰂魚涌英皇道1065號東達中心1305室
電話：2564 7511
傳真：2565 5539
電郵：info@wanlibk.com
網址：http://www.wanlibk.com
　　　http://www.facebook.com/wanlibk

發行者
香港聯合書刊物流有限公司
香港新界大埔汀麗路36號
中華商務印刷大廈3字樓
電話：2150 2100
傳真：2407 3062
電郵：info@suplogistics.com.hk

承印者
美雅印刷製本有限公司

出版日期
二零一八年五月第一次印刷

萬里機構

萬里 Facebook